Life Cycle of a

Silkworm

Ron Fridell
and
Patricia Walsh

Heinemann
LIBRARY

www.heinemann.co.uk/library

Visit our website to find out more information about Heinemann Library books.

To order:

 Phone 44 (0) 1865 888066

 Send a fax to 44 (0) 1865 314091

 Visit the Heinemann Bookshop at www.heinemann.co.uk/library to browse our catalogue and order online.

First published in Great Britain by Heinemann Library, Halley Court, Jordan Hill, Oxford OX2 8EJ
a division of Reed Educational and Professional Publishing Ltd.
Heinemann is a registered trademark of Reed Educational & Professional Publishing Ltd.

OXFORD   MELBOURNE   AUCKLAND   JOHANNESBURG   BLANTYRE   GABORONE
IBADAN   PORTSMOUTH (NH) USA   CHICAGO

Designed by Wilkinson Design
Illustrated by David Westerfield
Originated by Dot Gradations
Printed by South China Printing in Hong Kong.

ISBN 0431 08462 9 (hardback)   ISBN 0431 08463 7 (paperback)
05 04 03 02   06 05 04 03 02
10 9 8 7 6 5 4 3   10 9 8 7 6 5 4 3 2 1

British Library Cataloguing in Publication Data

Fridell, Ron
Life cycle of a silkworm
1. Silkworm - Life cycle
I. Title   II. Walsh, Patricia
595.7'8

Acknowledgements

The Publisher would like to thank the following for permission to reproduce photographs:
Em Ahart pp.12, 21; Black Star Publishing/Scott Rutherford/PictureQuest p.25; Bruce Coleman Inc./E. R. Degginger pp.7, 13, 26/Picturequest, /Pam Taylor p.10; Corbis/Gallo Images /Anthony Bannister pp.14, 15, 28, 29, /Wolfgang Kaehler p.24, /Charles and Josette Lenars p.23, /Stephanie Maze p.11, /Robert Pickett pp.20, 29; Dwight Kuhn pp.6, 8, 16, 19, 28; National Geographic Society/Stephanie Maze p.5; Photo Researchers Inc./S. Nagendra p.27; SPL/Photo Researchers Inc/Pascal Goetgheluck pp.16, 29; Stock, Boston/PictureQuest/Cary Wolinsky p.22; University of Nebraska/James Kalisch pp.4, 9, 17, 18, 29.

Cover photograph reproduced with the permission of Dwight Kuhn.

Special thanks to the University of Nebraska, Department of Entomology and Mr. James Kalisch.

Every effort has been made to contact copyright holders of any material reproduced in this book. Any omissions will be rectified in subsequent printings if notice is given to the Publisher.

Some words are shown in bold, **like this**. You can find out what they mean by looking in the Glossary.

Contents

Meet the silkworms

A silkworm is an **insect**. It is a worm, not a caterpillar. It looks as if it has many legs, but only six are true legs. The others are for clinging to plants.

1 day

3 weeks

5 weeks

Many years ago, silkworms lived in the wild. Today they live only on silk farms. Silkworms are **domesticated**. Silk farmers **raise** them to get the silk thread used to make silk cloth.

8 weeks

10 weeks

11 weeks

Egg 1 day

The silkworm begins life in a tiny egg. The egg is one of about 300 sticky, yellow eggs laid by the **female** silk **moth**.

1 day

3 weeks

5 weeks

The egg needs to be cold for a few weeks. Then the egg is warmed up and it turns a dark colour. The warm egg **hatches** in about ten days.

8 weeks

10 weeks

11 weeks

Hatching

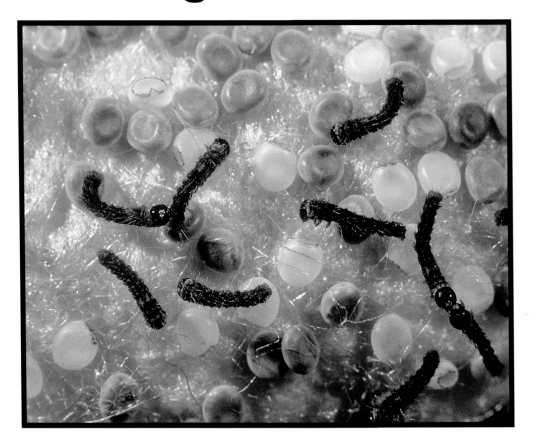

When the tiny silkworm **larva** is ready to **hatch**, it bites a hole in the egg. Then it wriggles out. At first the newly hatched larva looks like a tiny black string.

1 day

3 weeks

5 weeks

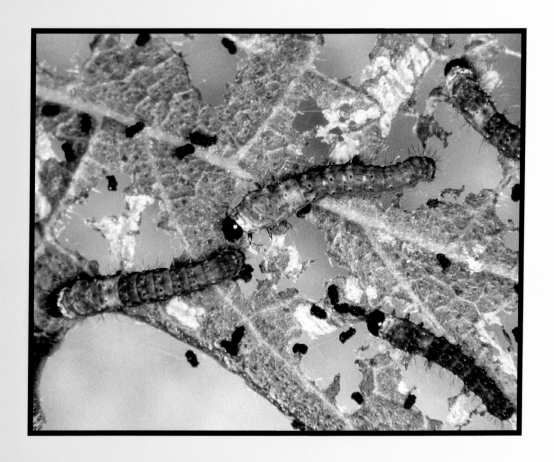

The young silkworm larva is too weak to crawl very far, but it is ready to eat. It needs to have its food nearby.

8 weeks

10 weeks

11 weeks

Larva

The silkworm **larva** only eats the leaves of the mulberry tree. In a few days, the larva will be stronger. It will be able to crawl from leaf to leaf.

1 day

3 weeks

5 weeks

The larva eats a lot of mulberry leaves. It does not drink water. It gets enough liquid from the leaves.

8 weeks

10 weeks

11 weeks

Moulting

The **larva**'s skin does not stretch as it grows. To get bigger, a silkworm must **moult**. The old skin splits. The silkworm wriggles out in its new skin.

1 day

3 weeks

5 weeks

The larva moults four times. After the fourth moult, the larva eats even more mulberry leaves than it did before. It grows to be about as long as your finger.

8 weeks

10 weeks

11 weeks

Spinning the cocoon   8 weeks

The **larva** is ready to spin a **cocoon**. It makes its cocoon from one long, sticky silk thread that comes from its mouth.

1 day

3 weeks

5 weeks

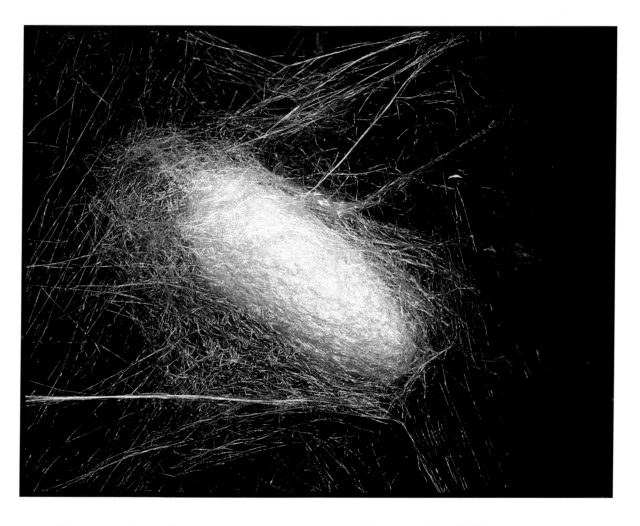

First the larva spins a silk web. Then it spins and spins for three days. It makes a silk cocoon around itself.

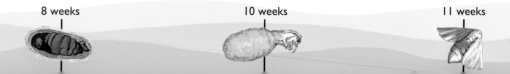

8 weeks   10 weeks   11 weeks

Pupa

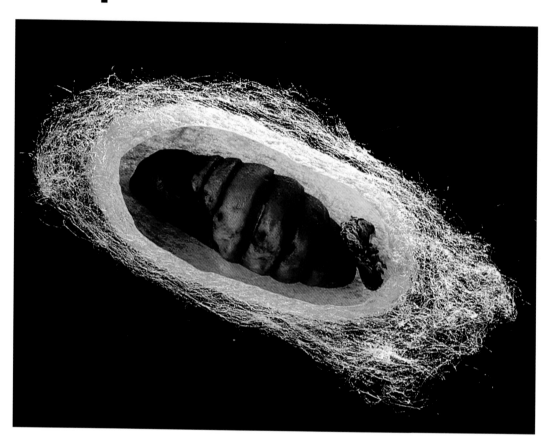

Inside the **cocoon**, the **larva** **moults** one last time. This time it changes into a brown **pupa** with a hard shell.

1 day

3 weeks

5 weeks

cocoon   moth

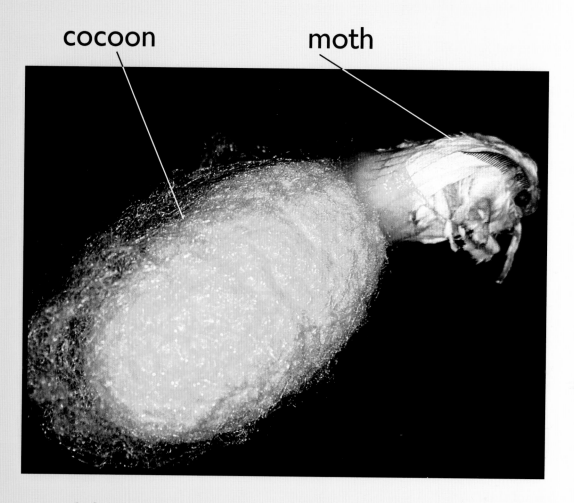

After two weeks, the shell splits,
and the pupa sheds this hard skin.
The pupa has changed into a
white adult **moth**.

8 weeks   10 weeks   11 weeks

Leaving the cocoon 10 weeks

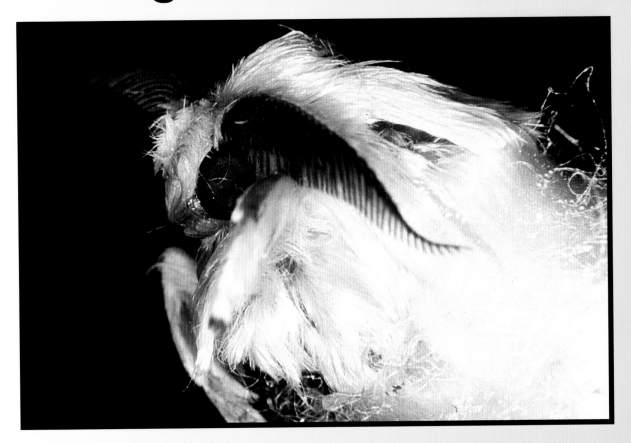

It is time for the silk **moth** to leave the **cocoon**. The moth spits out a special liquid. The liquid wets the inside of the cocoon and makes a hole in the strong silk.

I day   3 weeks

5 weeks

antennae   eye   wing

The silk moth pulls itself through the hole. It has wings, large eyes and feathery **antennae**. In about an hour, its damp wings unfold and dry.

19

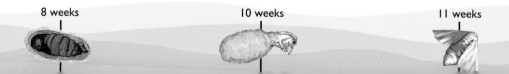

8 weeks   10 weeks   11 weeks

Silk moth

The silk **moth** has six legs and two **antennae**. It also has four wings, but it cannot fly. It only flutters and hops.

1 day

3 weeks

5 weeks

For the next few days, the silk moth does not eat or drink anything.

8 weeks

10 weeks

11 weeks

Silk from silkworms

Silk farmers **raise** silkworms for their **cocoons**. The silk thread of the cocoons is spun and woven into silk cloth.

1 day

3 weeks

5 weeks

On silk farms, most **pupas** never change into **moths**. If the moths came out of the cocoons, there would be holes in them. You cannot use these cocoons for making silk thread.

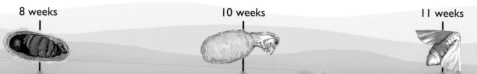

8 weeks   10 weeks   11 weeks

Making silk cloth

The long, white silk thread from the **cocoon** is as thin as a spider's web. It is unwound from the cocoon. Machines twist the threads together to make one **strand** of silk.

I day

3 weeks

5 weeks

The strands are woven into silk cloth. The cloth can be **dyed** any colour to make beautiful clothing. Silk clothes like these have been made in Japan for centuries.

8 weeks

10 weeks

11 weeks

Mating

The **female moth** is bigger than the **male**. Soon after coming out of the **cocoon**, she gives off a **scent** to help the male find her. Then they **mate.**

1 day

3 weeks

5 weeks

After mating, the male moth dies. The female moth lays her eggs a few hours after mating. Then she dies, too.

8 weeks

10 weeks

11 weeks

Life cycle

1 Egg

2 Young larva

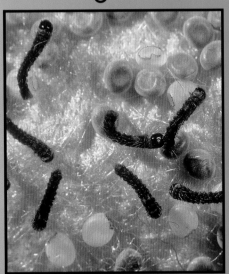

3 Larva

4 Cocoon

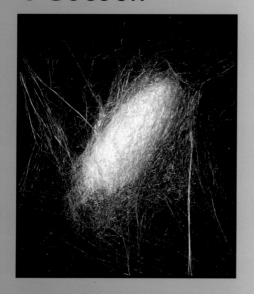

5 Pupa

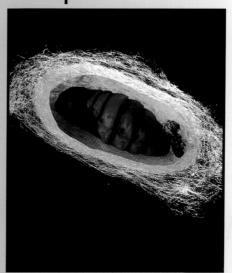

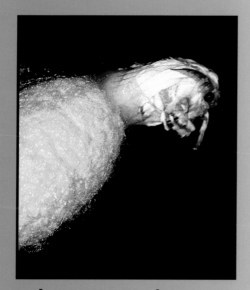

6 Leaving the cocoon

7 Silk moth

Fact file

The silkworm **larva** moves its head in a figure-of-eight pattern as it spins its **cocoon.**

It takes 110 cocoons to make enough silk for a silk tie. It takes 630 cocoons to make enough silk to make a silk shirt.

Other kinds of **moths** can fly to escape **predators**, but a silkworm moth has no predators and does not fly.

A single thread of silk from a cocoon can be as long as seventeen football fields laid end to end.

A single thread of silk is very strong – stronger than a thread of some types of steel (a very strong metal).

Glossary

antenna long, thin feeler on an insect's head. You say two, or many, antennae.

cocoon silk case that protects the pupa inside it

domesticated cared for by people

dye change the colour of fabric

female girl or woman

hatch come out of an egg

insect small animal that has six legs, a body with three parts, and wings

larva young stage of an insect's life when it eats a lot and grows quickly

male boy or man

mate when a male and female come together to make babies

moth insect with a thick body and four wings

moult shed the outer skin to allow an insect to grow

predator animal that eats other animals

pupa resting stage in an insect's life between when it is a larva and when it is an adult

raise care for an animal or plant until it is fully grown

scent odour or smell

strand thread or string

Index